A Guide to Archiving of Electronic Records

1st Edition
March 2014

A Guide to Archiving of Electronic Records

The Scientific Archivists Group (SAG) formed a Working Party to develop guidance to assist in the interpretation and application of the regulations, and to recommend best practice for archiving electronic records. This Guidance Document was produced by that Working Party drawing upon the collective experience of the contributors.

The following members of the Scientific Archivists Group participated in the Working Party that produced this guidance. Without their expertise and commitment, this guidance could not have been produced.

Authors:

Tim Stiles (Chair)	Qualogy Ltd
Julia Lawrence	Unilever, SEAC
Neil Gow	UCB
Eldin Rammell	Rammell Consulting Ltd
Gary Johnston	Audata Ltd
Russell Joyce	Heath Barrowcliff Consulting Ltd

First Printing: 2014

ISBN 978-0-9557659-2-6

Scientific Archivists Group Limited
Registered Office Berkeley Townsend, Hunter House, 150 Hutton Road, Shenfield, Essex, CM15 8NL, UK

www.sagroup.org.uk

Contents

1.   Introduction

The retention and archiving of study materials and process records, raw data and source data, is a critical part of compliance with both Good Laboratory Practice (GLP) and Good Clinical Practice (GCP).

The guidance contained within this document is also relevant and applicable to other areas of regulatory compliance such as Good Manufacturing Practice (GMP), Pharmacovigilance (PV), legislative and corporate financial requirements, and to other business critical records.

The maintenance and retention of such records provides the means by which a study, trial, process or procedure can be reconstructed and thus enabling the information and results generated to be verified.

Requirements for the operation of an archive and the archiving process for electronic records are no different to the requirement of physical records which are required to be retained for regulatory or business requirements. However, due to the variety and fragility of electronic records some additional features are required. The purpose of this document is to identify and discuss these features and provide guidance on how these challenges can be met.

2.   Definition of terms

Archives: The facilities and supporting resources necessary for the secure retention, maintenance and retrieval of materials accumulated by an organisation.

Archivist: An individual designated by management to be accountable for the management of the archive, i.e. for the operations and procedures for archiving.

Electronic archive: The designated repository in which electronic records are retained for their long term preservation.

Electronic Document and Records Management System (EDRMS): A computerised system designed to capture electronic documents and electronic records and manage them in a controlled way.

Electronic record: Information recorded in electronic form that requires a computerised system to access or process.

Information Technology staff: Designated personnel given responsibility for providing technical support to staff using a computerised system.

Ingestion: The process that accepts electronic records for archiving.

Material(s): A collective term given to all documents, data and records which need to be retained for regulatory purposes. This includes, but is not limited to; raw data, source data, process records and non-study/trial specific records necessary for the reconstruction of the study/trial. This includes records generated and maintained in electronic form.

Metadata: Information associated with data that provides context and understanding, i.e. data about data. Most commonly this is data that describes the structure, data elements, inter-relationships and other characteristics of electronic records.

Migration: The transfer of electronic records from a source format or system to a target format or system.

Preservation: Ensuring the electronic records held in the archive remain accessible through the application of appropriate preservation policy and processes. In the context of electronic archiving this means maintenance of the authenticity and integrity of electronic records.

Refreshment: Procedures to maintain the physical viability and continued readability of the storage media.

Storage media: The different physical materials on which information may be recorded. Examples include paper, photographic film, magnetic media, microforms and optical devices, local and remote servers.

3.   Scope

The approach taken in this guidance is from the perspective of Archivists and their role and responsibilities in the retention of electronic records.

This approach is intended to provide guidance to the Archivist on the regulatory expectation of their role in the retention of electronic records and to identify some of the challenges this expectation presents.

The guidance is not intended to be restricted to regulatory records. The principles it describes can equally be applied to non regulatory records such as financial and personnel records

4.   Regulatory Perspective

The principles and compliance aspects related to the archiving of electronic records are no different to that of paper. Clearly the physical aspects of archiving electronic records are very different to paper and will therefore require different procedures.

OECD Principles of GLP defines Test Facility Management as being responsible for the long term protection and retention of study and facility records. They achieve this by the provision of suitable archive facilities for the storage and protection of the records and the appointment of an Archivist to manage the day to day operation of the archive facility.

In the ICH GCP Guidelines the role of the Archivist is not defined but the retention of "essential documents" is clearly expressed via the maintenance of the trial master file and investigator site files. The EC Directive on Good Clinical Practice requires the appointment of an Archivist. The expectation of GCP inspectors for the operation of an archive of clinical source data and records by Sponsors would appear to be no different to those within GLP.

Section 4.1 of EU GMP states that "Secure controls must be in place to ensure the integrity of the record throughout the retention period and validated where appropriate".

Much of the available regulatory guidance is from a GLP perspective as these regulatory requirements have been in place and subject to audit for over thirty years. For this reason much of the guidance within this document has its foundation in our experiences from a GLP perspective. Guidance on the role of the Archivist has been published by the OECD in their guidance Monograph No. 15, entitled 'Establishment and Control of Archives that Operate in Compliance with the Principles of GLP'.

However, one overriding principle in all regulations is that for records to be considered to have been archived such records should come under the control of an appointed Archivist.

We should also be cognisant of the US regulation on "Electronic Records, Electronic Signatures", US Code of Federal Regulations - Title 21 Part 11 (21CFR11). These regulations are focused on the generation and use of electronic records that are required under US Food and Drug Administration (FDA) regulations, and the maintenance of their authenticity and validity. It also includes a section covering electronic signatures. Compliance with the regulation is mandated if the records are required by a US FDA regulation.

Many of the requirements of 21CFR11 remain valid despite the changes in technology since its first publication. This guidance should not conflict with the expectations and requirements of these regulations.

4.1   Application of Regulations to archiving of electronic records

The decision to archive records in electronic form must be made prior to their creation as such decisions have ramifications on the validation of the system and critical implications for record generation and archiving. Such decisions will be influenced by a number of factors, but its form (physical or electronic) must be decided prior to the point of record generation and such data definition documented in relevant SOPs.

In some cases data such as output from scientific instruments are collected electronically printed, reviewed and approved, defining this output as the raw data. The data can be processed electronically to the reporting stage and indeed saved electronically, with the paper print out defined as the raw data and archived. The final report will be audited against the paper raw data to confirm the accuracy of reporting.

The equipment used in the generation, processing and retention of electronic records should be tested to demonstrate the reliability and accuracy of the system and its ability to capture such data and to retain that data in its original form. The process of computer systems validation is not a topic that will be discussed in this document. Requirements for the validation of computerised systems can be found elsewhere, however the need for systems validation and demonstrating equipment is fit for purpose is unequivocal and equally applies to systems used in the capture, processing and archiving of electronic records. The principles defined by FDA for Electronic records and Electronic signatures (21CFR11) will similarly need to be applied in the process of archiving electronic records.

As with paper records, alterations or corrections must be appropriately recorded without obliterating the original records. This information is usually recorded in the audit trail.

Metadata also need to be retained. The regulations applicable to the electronic records also apply to the metadata.

5. Roles and Responsibilities

5.1 Role of the Archivist

The term "Archivist" is used to identify the individual appointed to take accountability for the operation of the archive and the management of archived records.

In an organisation that does not actually operate its own archive facility there must still be an Archivist who provides the controls necessary to ensure such records are protected, controlled and managed to maintain its integrity and validity and accessible to Regulatory Authorities when required.

The Archivist is responsible for:

- Ensuring that records submitted for archiving corresponds to that described in transmittal documentation (chain of custody). Archive staff should verify that the records received correspond with the information provided by the depositor. There should be a documented acceptance of the records into the archive by the archive staff.

- Ensuring that orderly storage and retrieval of records is facilitated (e.g. by means of indexing).

- Controlling access to the archive and the archive records.

- Ensuring that movement of archived records is properly controlled and documented.

- Ensuring preservation of records held in the archive.

The ingestion of records into some electronic archive systems can be automated, however, the principles as defined above should still apply. The transfer of responsibility from the depositor to the Archivist should not be taken for granted given the automation involved. The Archivist should be aware of the point at which they become responsible for the electronic records and when access controls to the records should be applied.

Job descriptions and SOPs, and where necessary contracts and policies, should be established that define how the role is performed, the electronic archive operates and the records are managed.

In larger archives there may be a need for several archive staff to effectively undertake the operation of the archive and the archiving of the records. These staff should work under the supervision of the designated Archivist.

The Archivist and archive staff should be appropriately trained in the operation of the electronic archive. Evidence of the training received should be documented.

5.2   Role of Management

In a GLP environment there are clearly defined responsibilities with regard to establishing an archive and the archiving process. For example, the Study Director is responsible for ensuring study material is archived promptly and completely at the end of the study.

In a GCP environment, the investigator and sponsor have responsibilities to protect trial records and patient confidentiality. The investigator site file is the responsibility of the investigator and this file should not be retained by the sponsor. The independence of the investigator site file should be maintained when such records are held electronically.

In a GMP environment, there are clear requirements detailed in the EudraLex Chapter 4 of EU GMP and in US CFR Title 21 Part 211 regulations listing the types of records which should be retained and archived. These include production records, analytical data, batch release records, investigations, deviations and change control records.

Management has the responsibility to provide appropriate facilities in which records are to be archived, and that supporting records such as staff training records, equipment maintenance records, audit records and process records should be retained and archived.

In the same way as in a paper environment Management provide the physical archive, the storage medium (shelving) and manage the environmental control of the facility (air conditioning, security systems etc), so in an electronic environment they will provide the computing infrastructure, the data servers and control systems (network access, backups etc.) to enable the operation of the electronic archive by IT.

5.3   Role of sponsors

For nonclinical and clinical research studies, the sponsors are ultimately responsible for ensuring the records that support the data submitted to regulatory authorities are retained and available when required, even if they remain under the management of a contractor. The condition under which such records are retained should ensure its integrity, validity and accessibility.

5.4   Role of Information Technology staff:

IT staff are responsible for:

- Maintaining and supporting computer software, hardware and IT infrastructure.
- Maintaining system configuration

- Performing routine backups of the system or overseeing automated backups at defined intervals.

- Providing technical support to the business managers and end-users.

- Undertaking technical aspects of the system validation.

All activities and processes carried out by IT staff should be appropriately documented in SOPs. IT staff must follow procedures agreed with the Archivist and/or Management and receive appropriate awareness training in GLP/GCP/GMP regulations when applicable.

Where an electronic archive process or system is contracted to a third party some of the above IT responsibilities may be retained in-house. The delegation of responsibilities must be documented.

6. Options for archiving electronic records

6.1 Archive Mechanisms

There are two primary mechanisms for archiving electronic materials. These are:

a. a defined archive area as part of a computerised system;
b. a designated electronic archive system or systems.

a) Defined archive area

Where a computerised system has its own archive area, this may be physically separated from non-archive records or there may be a virtual separation. For example, records may be physically separated by storing them in a separate file system , or virtually separated using metadata to denote archived records. In either case, archived records must be locked such that they cannot be altered or deleted without detection and must be under the control of the nominated Archivist.

b) Designated archive system

Where a designated electronic archive system is used, it is permissible to have a single archive application for the management of all archived electronic records or to have a number of archive systems that are designed to work with specified feeder applications. Where an organisation has a limited number of computer applications and/or a limited number of record types to be archived, a single archive system often is the preferred approach.

It is essential that there is no ambiguity where archived records are being retained and managed. This should be appropriately documented, for example in SOPs or a system specification.

The archive may have a single interface for the end-users that enables the depositor to transfer records from any record-keeping system into the appropriate archive system or a customised interface for each application. Many off-the-shelf archive systems provide full integration with a range of systems such that the end-user is provided with an archive option from within the desktop application.

6.2   Outsourcing IT

In order to maintain flexibility within their infrastructure, organisations may outsource some or all of the IT systems and services including archiving. Archiving regulations and standards do not preclude the use of third parties. However there are additional factors that need to be considered, including:

- the storage location of data must comply with Data Protection legislation and any Safe Harbour agreements;

- disaster recovery processes must be in place to ensure the data are secure;

- a process for transfer of records following contract termination must be in place;

- the ability of the network connection to efficiently transfer large volumes of data must be assessed for performance and reliability.

6.3   Use of portable media

An alternative approach may be to migrate electronic records from the source application to portable media. Examples include magnetic tape, DVD/CD optical disk, external hard drives or USB flash drives. These can be placed in a physical archive and managed in the same manner as other hard-copy records.

When archiving electronic records in this way the same archiving principles that apply to other physical records - such as deposits, loans and retrievals - would also apply.

We strongly advise that this approach is not used for archiving of electronic records with long retention times, for the following reasons:

- the longevity of the archive storage medium is uncertain. Whilst some manufacturers claim that their media will last in excess of 50 years, various accelerated tests have demonstrated potential failure within a 5 year period or less.

- the issues associated with data refreshing and migration will not be addressed therefore leading to loss of data through media degradation.

- the ability to read the storage media in the future may be compromised due to hardware obsolescence. Examples of media which are now approaching obsolescence are floppy discs, DAT tapes, compact cassettes.

- poor search and retrieval capability for records distributed across portable media.

To alleviate the issues caused by obsolescence and degradation this approach would require regular testing of the storage media used and the regular migration of data from old media to newer media, validating each migration.

Archival tape is an acceptable archive storage medium if it is being used in a tape library and is being appropriately managed.

6.4   Source of Electronic Records

The requirements for electronic archiving and digital preservation should be considered as a key aspect of the procurement or development of any computerised system that creates, captures or manages electronic records.

This not only includes electronic records management systems but also Laboratory Information Management Systems (LIMS), computerised laboratory equipment (e.g. HPLC), electronic quality management systems and a wide range of other business applications. All of these systems hold electronic records that may have to be archived in electronic format. Therefore due consideration must be given to the long term retention requirements for electronic records, particularly in relation to accessibility, readability, and possibly any future processing needs when procuring or developing of such systems.

Technical and functional requirement specifications for such computer systems should include requirements for the long-term storage, preservation, management and retrieval of their electronic records. The set of requirements should include all of the issues described within this guidance to ensure that electronic records from those systems can be archived in a manner that is compliant with all applicable requirements and industry best practice.

6.5   Outsourced record capture

Where activities are outsourced to third parties, e.g. bioanalytical analysis, data management or CRO, plans should be in place to ensure that arrangements for the archiving of any electronic records will permit continuation/preservation of the system on which to access the electronic records. Alternatively the electronic records, including the meta data, should be preserved in a format that does not rely on the source system remaining available.

7. Electronic Archive Processes

This section discusses the factors that should be considered when developing a process for archiving electronic records and the archiving issues that should be considered when acquiring or developing a computerised system that captures, creates or manages electronic records.

These principles should equally apply to off-the-shelf systems and to those which are custom built to the organisation's specification. The same principles may be applied if an existing system or file storage mechanism is being extended in scope to become a designated archive.

Additional guidance may be found in the following reference materials:

- ISO/TC 46/SC 11 Digital Records Preservation – Where to start guide;
- Open Archival Information System (OAIS) Reference Model (ISO 14721:2012); and
- Audit and Certification of Trustworthy Digital Repositories (TDR) checklist (ISO16363:2012).

The process must demonstrate that the systems and procedures followed ensure the usability, reliability, integrity, authenticity and accuracy of the data produced and that the systems operated are secure to protect the records from loss or uncontrolled alteration throughout the life of the record.

7.1 Selection, capture and processing

Records may be captured from standalone systems such as an automated temperature monitoring system for fridges and freezers in a relatively straight forward manner. A file of individual values can be transferred to an appropriate electronic storage medium.

More complex systems may include:

- process data generated from different analytical platforms
- composite information from other systems, for example databases.

In either scenario the system that generates electronic records should be assessed individually in order to determine the archive requirements and archive process. A specification should be produced for each system which describes the movement of the electronic records from the point of capture/generation through to the point of archiving. The location of the electronic records throughout the process should be clearly stated.

At every stage the electronic records should be protected by access controls and audit trails. Prior to archiving, any permitted alterations or amendments to the electronic records must be traceable by recording who made the change, when the change was made, what has been changed and why it has been changed. Examples may include the conversion of material to a human readable format or the addition of further metadata to assist management of archived content. The original record may also need to be retained.

At the end of this stage the record, including all metadata, must become fixed.

7.2   Ingestion

This stage commences once the electronic record is ready for archiving. The process for presenting electronic records for archiving should include the following aspects:

- Confirmation between the depositor and the Archivist as to the electronic records to be archived.

- The timing of the transfer of electronic records to the archive.

- The identification of the records being archived. The metadata provided for the electronic records being archived should be such they can be easily identified, located and retrieved.

- Verification by the Archivist of the successful and complete transfer of the electronic records will depend upon the type of system being used. For simple systems, counting the size and number of files transferred may be sufficient but for complex systems a more rigorous process of data sampling may be required.

When utilizing automated electronic archive processes the principles as defined above should still apply.   These are usually defined and applied during set up of the system.

In some situations the assistance of IT staff will be necessary to complete the archiving process.

At this point, the records are deemed to have been archived and have passed to the control of the Archivist.

7.3 Storage

Electronic records should be archived so as to ensure their reliability, authenticity, integrity and usability at all times. Storage of the archived materials should be aligned with the records preservation plan. Electronic archives should be backed up.

The Archivist may delegate responsibility for the installation, operation and performance of the storage system to IT staff or a contractor. A process for managing and maintaining the electronic archive system should be agreed with the Archivist.

7.4 Documentation

Policies and SOPs that describe the archiving process should be in place; these will include, but not necessarily be limited to, the following:

- Definition and description of what constitutes the archive.
- Description of acceptable file formats.
- The conditions under which the electronic records is stored and any monitoring of those conditions.
- Procedures to ensure the integrity of the archive storage and the records archived. This must include periodic checks on the conditions of the records archived to ensure deterioration of the storage media or bit level degradation has not occurred.
- Procedures for the receipt and checking of electronic records to be archived.
- Maintenance of archived records, including the refreshing and migration of electronic records.

- Definition of required metadata to facilitate searching of the archived electronic records to ensure timely and accurate retrieval.

- Documented retention period for records archived.

- Security of the archive systems and the electronic records retained, including access controls.

- Procedures for the backup of electronic records.

- Procedures to define preservation actions to ensure continued access to the electronic records throughout their retention period. This should be described in a preservation plan.

- Access control

- Responsibilities for the operation of an archive and the role of the Archivist.

- Disaster recovery policy

7.5   Retention and destruction

Retention periods apply regardless of the format of the record. Records should be retained for the period(s) specified in the organisation's records retention schedule. Management should be responsible for preparing a records retention schedule for the different types of electronic records, taking into account the business, regulatory and legislative requirements.

Electronic records should be organised and stored according to the classification system established within the organisation in order to manage groups of records in the future for possible reuse, transfer to a business partner or subsequent destruction. It is strongly recommended that the classification of records takes place at the time of creation.

Organisations should have a records destruction policy and this should include electronic records. Organisations should retain a record of destruction that list the records destroyed in order to provide evidence of compliance with the records retention policy. Records of destruction should show that destruction was conducted in accordance with all applicable regulations.

For electronic records, the destruction policy should stipulate whether destruction relates to removal of the "pointer" to the files or to actual electronic records. In any event, there should be a mechanism to ensure that electronic records identified for destruction are not recoverable. Since the deletion of "pointers" does not physically destroy the files, additional steps such as overwriting the data may be necessary. Where archived records are encrypted, destruction of the encryption key may be an acceptable means of destroying the records. If a portable digital medium is used (eg CDs, DVDs, USB devices, hard drives) appropriate means should be taken to render such media unusable.

Consideration should also be given to destroying any copies of electronic records due for destruction that may be held on back-up tapes and whether or not to destroy these records in line with the regular back-up tape cycle.

7.6   Access and security

In the electronic environment, "access" is defined as taking a copy of an electronic record. Simply consulting a read-only copy is "viewing"

Processes should be in place to ensure that access to the archived records in the electronic archive is controlled by the Archivist; this may require the use of a password controlled log-on or access levels restricted according to the records classification. The electronic archive should maintain an audit trail of all who have access and have accessed archived materials through log-on or other suitable control mechanisms.

Electronic records that have been archived may, on occasion, need to be viewed. The ability to view archived records should be restricted to individuals with a genuine business need. It should not be necessary to maintain a log of records viewed in this way.

Processes should be implemented to ensure that the archived records cannot be altered, moved or destroyed without appropriate authorisation. Any approved action of this nature should be appropriately documented.

If the electronic records are archived on portable media the same principles should apply. It is strongly recommended that copies of the information are made if required and the original medium should remain in the archive to prevent the risk of loss or damage to the original records. This is not a loan but providing a copy of the information.

Circumstances where electronic records are permanently removed from an archive such as:

- change of the ownership of the records.
- move from one archive location to another.

The release of records from the archive, under the circumstances above, should be documented and approved by Management. Following transfer of electronic material from the archive, the deletion procedures described above must be applied.

8.   Preservation strategy

Retention of electronic records presents challenges not present in the archiving of physical materials. These challenges and the risks they pose are discussed below. Archived electronic records should be protected from potential loss caused by these risks.

8.1   Media Degradation

The media on which electronic records are stored is inherently unstable and without suitable storage conditions and management can deteriorate quickly even though it may not appear to be damaged.

Processes should be in place to ensure that the media on which the archived electronic records are stored does not degrade thereby preventing the archived records from being read. At the time of selecting the storage media to be used consideration should be given to the length of time the electronic record is to be retained. Archived electronic records must be migrated to an alternative medium prior to the degradation of the media. The ability to read the media must be regularly checked. Any such migration should be documented and validated.

Individual bits within documents can become corrupted thus rendering the file inaccessible. This is known as Bit level degradation and can be caused by a number of mechanisms such as file transfer, media depredation and environmental impacts (temp, humidity etc). The bit level integrity of each file should be checked periodically through the use of checksums. ISO 16363 provides guidance on the use of such checksums.

8.2   Media Obsolescence

Processes should be in place to ensure that the media on which the archived electronic records are stored does not become unreadable because the hardware and software required to read it becomes unavailable.   This might be achieved by migrating the archived records from one storage media to another, for example from a floppy disc to a hard drive.   Any such migration should be documented and validated.

8.3   Hardware Obsolescence

Many of the electronic records archived have a retention period in excess of the typical life expectancy of most electronic hardware. This means that the hardware used to archive electronic records and the hardware that is needed to access the electronic records often becomes redundant or obsolete many years before the electronic records themselves can be destroyed.

Archived electronic records should be protected from loss due to hardware obsolescence.   Processes should be in place to migrate the electronic records to ensure accessibility by current and future hardware.

8.4   Software Obsolescence

The software development life cycle may result in electronic records created by older versions of software not being fully compatible with newer versions. In addition software may no longer be available thereby rendering access to electronic records less likely. It is important therefore that a digital preservation strategy identifies how to mitigate for software obsolescence and the impact on previously generated electronic records.

This might be achieved by a regime of monitoring the archived electronic records and then migrating from one software system to another, for example from one EDRMS to another, migrating from one file format to another, for example MS Word 2 to MS Word 10, or conversion to another file format, for example MS Word to PDFA. Any such migration should be documented and validated.

8.5   Digital Preservation Strategy

Electronic records simply left or locked away on a computer or computerised system does not constitute an electronic archive. An electronic archive must be carefully planned and a digital preservation strategy implemented.   Without a preservation plan there is no electronic archive.

To ensure that the archived electronic records are retrievable and readable over the whole of their retention period, the preservation plan should describe how the issues described earlier are addressed.

There is a balance to be achieved between the amount of structure that is retained in archived electronic records against the ease of future maintenance and readability of the records. For example, data stored in an unstructured text format which conforms to a commonly adopted standard can be expected to be easily readable without any specialist software for many years, but there would be limited opportunity for reanalysis or automated review.

Electronic records stored in a highly structured format such as a complex database could be rapidly searched or analysed if retained in this format.   In order to do so, some suitable computer software (and also potentially the hardware and the operating system) would have to be available for as long as the records are required, and it may become difficult just to view the records without the appropriate tools.

Therefore, it is recommended that a digital record preservation strategy be developed which assesses the likely business, regulatory and legal needs for the electronic records throughout their retention period. Where there is likely to be a future need to process (e.g. reuse or reanalyse) the records, a format should be chosen that will permit future processing.

It should be highlighted however, that these formats are typically less stable over time and there is therefore a higher likelihood that the electronic records will need to be migrated to newer file formats for records that have lengthy retention times. In contrast, if there is unlikely to be a need to reprocess the content, a formatted record file type can be chosen that is more sustainable over time. This allows the record content to be displayed in a reliable manner but is unlikely to allow the content to be processed again. Examples of this type of file format include PDF/A, PDF/X and XML.

8.6   Future proofing

Consideration should be given to the possibility that the electronic archive system software could itself become obsolete over time and require updating or replacement. The electronic archive software may also impose its own limitations, if later versions of the software are not able to support the archived file formats and their versions. It is wise to consider this issue and discuss potential migration options with vendors at the time of selecting an electronic archiving solution.

As with any computerised system, consideration should be given to the ultimate retirement of an electronic archive system, or how to respond if it becomes necessary to abandon use of a system. For example, if the vendor ceases to trade and can no longer support the system, this could became an acute issue, particularly for cloud-based systems because of the potential risk of immediate loss of access and difficulty of reinstatement. A strategy should be in place at the time of purchase to ensure that options exist to transfer data to an alternative archive system employing a validated process if and when it becomes necessary to do so.

9.   Summary

The challenges of archiving electronic records and the role of the Archivist in this endeavour should not be underestimated nor taken for granted. Whilst the regulatory principles of archiving electronic records are no different to those of paper records, the arrangements for and management of an electronic archive present new and different challenges.

The definition of what constitutes raw or source data when an observation is captured electronically is the point of creation of that record. The manner in which that record is controlled, migrated and managed is key to ensuring the electronic record and any associated meta data, when presented for archive retention, fulfil the regulatory definition of raw or source data.

The operation of a regulatory electronic archive remains the role of the Archivist, appointed by Management, who is suitably trained and qualified for the role.

The operation of an electronic archive presents a number of different challenges for the Archivist. Ensuring the maintenance of the media on which the record is held remains fit for purpose, the ability to access hardware and software to enable the reading of that media, the management and access control of the archived electronic records and eventual destruction of the electronic record will represent new knowledge and skills   for many Archivists.

It is hoped that this guidance document, whilst highlighting many of the challenges and threats to archiving electronic records, will provide greater understanding for and guidance to the Archivist in fulfilling their role and responsibility.

10. Resources

Association for Clinical Data Management (ACDM), Computer Systems Validation in Clinical Research - A Practical Guide (Edition 2, 2004)

Digital Preservation Council's Technology Watch on File Formats.

EU Commission Directive 2005/28/EC of 8 April 2005, laying down principles and detailed guidelines for good clinical practice as regards investigational medicinal products for human use, as well as the requirements for authorisation of the manufacturing or importation of such products

EudraLex The Rules Governing Medicinal Products in the EU, Volume 4 GMP, Chapter 4: Documentation

FDA Draft Guidance for Industry: Electronic Source Documents, December 2010, UCM239052

FDA: 21 CFR Part 11; Electronic Records; Electronic Signatures Rule.

FDA: 21 CFR Part 211 Current Good Manufacturing Process for Finished Pharmaceuticals

GAMP 4 Recommended Environmental Conditions for Storing Various Recording Media

GAMP Good Practice Guide: Electronic Data Archiving, ISPE, July 2007

ICH Harmonised Tripartite Guideline for Good Clinical Practice E6(R1), CPMP/ICH/135/95/Step5, 1996

ISO 11799: 2003(E) Information and documentation — Document storage requirements for archive and library materials.

ISO 14721:2012 Space data and information transfer systems -- Open archival information system (OAIS) -- Reference model

ISO 15489-1:2001 Information and documentation -- Records management

ISO 16363:2012 Space data and information transfer systems -- Audit and certification of trustworthy digital repositories

ISO 19005-1:2005 and ISO 19005-5:2010 Document management. Electronic document file format for long-term preservation (the PDF/A standard)

ISO/TC 46/SC 11 Digital Records Protection: Where to start guide. Understanding the issues specific to the preservation of digital records. Development of a preservation plan. Safeguarding digital records over time with confidence.

MHRA UK Medicines and Healthcare products Regulatory Agency (MHRA). Good Laboratory Practice - Guidance on Archiving. March 2006.

OECD GLP Principles. OECD Series on principles of Good Laboratory Practice and compliance monitoring, Number 1. ENV/MC/CHEM(98)17

OECD Series on Principles of Good Laboratory Practice and Compliance Monitoring no. 10: GLP Consensus document "The application of the Principles of GLP to Computerised Systems"

OECD Series on Principles of Good Laboratory Practice and Compliance Monitoring No. 15: Establishment and Control of Archives that Operate in Compliance with the Principles of GLP (published in 2007).

SAG Scientific Archivists Group. Guidance on the archiving of Good Clinical Practice material.

www.ingramcontent.com/pod-product-compliance
Lightning Source LLC
Chambersburg PA
CBHW021916190326
41519CB00008B/809